Friction in Transformation

A Design-First Approach to Organizational Change

By

Nathan Christopher Pease

Friction in Transformation by Nathan Pease
Published by Pease Patch Publishing

ISBN 13: 978-1-7370939-1-6
For information regarding permission, write to:
peasepatchpublishing@gmail.com

To my loving wife Heather, you are the real author in the family.

Everything I do is for my family. Heather and our three children, Jarek, Irish, and Draper, are my motivation for all good I do with my life. They force me to look through alternative lenses, to constantly challenge how I view the world and how things work.

Introduction: Why Change Feels Impossible (And Why It Isn't)

I have always wondered why change was so extremely difficult in organizations.

Early in my career, I believed the answer was simple. People resisted change because they didn't understand it. They lacked awareness of what was possible. They were comfortable, complacent, or afraid of technology. If we could just educate them, challenge them, push them harder, things would move.

So that's what I did.

I challenged teams to do better. To try harder. To learn more. To execute faster. I pushed people to break through the ceilings holding them down. I challenged leadership to make decisions, remove roadblocks, and "get things done." I believed progress was mostly a matter of effort and willpower.

And for a while, that belief seemed to work. Projects moved. Systems launched. Metrics improved—at least on paper.

But over time, something didn't sit right.

I watched talented, motivated people burn out. I watched leaders with good intentions stall. I watched smart initiatives collapse under their own weight. I watched organizations invest millions into technology, consultants, and transformation programs—only to quietly revert back to old habits months later.

The more success I had, the more uncomfortable I became with how fragile it all felt.

Today, my position is very different.

I've seen behind the red tape. I've sat in rooms where decisions are far more complex than they appear from the outside. I understand now that at every level of an organization, decisions are difficult—not because leaders don't care, but because it is impossible to make everyone happy.

Every decision has a cost. Every choice carries risk. And every move carries professional consequences that extend beyond the decision itself.

From shareholders to frontline operators, from customers to regulators, from quarterly targets to long-term survival—leaders are constantly balancing competing pressures. Add relentless global competition, shrinking margins, higher expectations, tighter labor markets, and more data than anyone can realistically process, and something interesting happens.

Transformation stops being the goal.

Survival becomes the priority.

Leaders spend their days fighting fires, navigating politics, managing risk, and keeping the machine running. Strategy turns into reaction. Vision gets buried under urgency. And the very people tasked with "driving change" are often the most constrained by the systems they operate within.

This is not a failure of leadership.

It is a failure of systems.

I've spent most of my career in technology—operational technology, digital systems, automation, data, and the messy intersection where people and machines meet. Not because I love technology, but because I learned early that technology exposes truth.

I'm not naturally drawn to screens or software. I'd rather be outdoors, chasing adventure, living unconventionally, finding space to breathe. But I've always picked up technology quickly. I'm not intimidated by it. I see it as a tool—one that can either amplify dysfunction or remove friction.

And over time, I realized something critical:

Technology doesn't fail organizations.
Organizations fail to design themselves to use technology well.

I've built systems. I've fixed broken ones. I've integrated platforms that were never meant to talk to each other. I've optimized processes that looked efficient but weren't effective. But the work I care most about has never been the tech itself.

It's the moment when a team suddenly "gets it."

When a system removes frustration instead of adding to it.
When someone's job becomes easier instead of harder.
When clarity replaces chaos.

That's when transformation actually happens.

Not when a new tool is installed.
Not when a dashboard goes live.
Not when a consultant declares victory.

Transformation happens when friction is removed from the system.

Over the years, I've watched countless organizations begin transformation with optimism and excitement—only to slowly lose momentum as reality sets in. Roadblocks appear. Dependencies surface. Resistance grows. The "culture isn't ready." Priorities shift. Budgets tighten. Leaders change roles.

Initiatives stall, not because they were bad ideas, but because the system they were introduced into was never designed to support them.

Culture gets blamed. People get blamed. Leadership gets blamed.

But culture is not the root cause. Culture is the outcome.

What I've learned—often the hard way—is that most transformation efforts fail long before execution ever begins. They fail at the moment we misdiagnose the problem.

We treat symptoms instead of systems.

We apply pressure where redesign is needed.
We demand behavior change where structural change is required.
We introduce tools where clarity is missing.
We ask people to "own" outcomes they have no control over.

And then we wonder why nothing sticks.

This book exists because I believe we can do better.

Not by trying harder.
Not by moving faster.
Not by adding more technology.

But by learning to see friction clearly—and then deliberately designing it out of the system.

Friction shows up everywhere:
In decision-making.
In accountability.
In handoffs.
In incentives.
In metrics.
In tools.
In meetings.
In well-meaning policies that made sense once but no longer do.

Friction is invisible when you're inside it. Normalized. Accepted. Explained away as "just how things work."

Until someone stops and asks a different question.

This book is not about digital transformation, operational excellence, or technology—at least not in the way those terms are usually discussed.

It's about learning how to identify friction, understand why it exists, and design systems that allow people to succeed without heroics.

It's about moving from effort-based leadership to system-based thinking.

It's about recognizing that most people are doing the best they can within the constraints they're given—and that real change begins when we redesign those constraints.

If you've ever felt frustrated by stalled initiatives…
If you've ever watched a good idea die quietly…
If you've ever wondered why capable teams struggle to execute…
If you've ever been told to "own" a result without the authority to influence it…

This book is for you.

Not because you're failing—but because the system might be.

And systems, unlike people, can be redesigned.

A Note on Reinforcement

You will see key principles repeated throughout this book. That is deliberate.

In manufacturing, stable performance comes from standard work, repetition, and reinforcement—not one-time instruction. The same is true in transformation. Friction does not disappear because it was identified once. It resurfaces in new forms, at different layers of the system.

Core concepts are revisited to reflect how real improvement works: diagnose, design, reinforce, repeat.

If a theme appears more than once, it is because sustainable change requires it.

Chapter 1: Why Transformation Feels Hard

When Progress Slows Without Resistance

Alex joined NorthRiver Foods as Director of Operations Technology with a clear mandate: help modernize operations in a company that knew it was behind but wasn't quite sure where to start.

NorthRiver was not anti-technology. Leadership talked openly about data, automation, and digital tools. Investments had been approved before. Vendors were familiar faces. Roadmaps existed—at least in slide form.

Yet the results were uneven.

Some initiatives stalled. Others launched but never scaled. A few quietly disappeared without ever being declared failures. When leaders discussed transformation progress, the language was careful and familiar.

"We need better alignment."
"There's resistance at the sites."
"We're not ready to move that fast."

Alex heard none of this as defiance. What he heard was hesitation—measured, professional, and rational.

Six months in, he realized the problem wasn't a lack of ambition or skill.

It was friction.

Month One: A System That Looked Coordinated—but Wasn't

From the outside, NorthRiver's structure appeared reasonable.

IT owned systems, security, and standards.
Operations owned performance.
Engineering owned equipment.
Finance owned capital and controls.
Operational Excellence sat in the middle, tasked with improving how everything worked together.

Alex reported into OpX (Operations Excellence), not IT. That choice alone sent mixed signals.

When Alex asked how technology priorities were set, he received different answers depending on who he asked. IT believed they were the intake point for all technology decisions. Operations believed they defined the need. Finance believed approval flowed through capital planning. OpX believed it was responsible for outcomes.

All of them were right.

And none of them owned the whole.

This didn't create conflict. It created drag.

Decisions required pre-alignment before alignment. Meetings were held to prepare for meetings where the real decision might eventually happen. Leaders spent time managing risk exposure rather than solving problems.

No one was resisting.
Everyone was compensating.

Month Two: Friction Shows Up in Behavior

By the second month, Alex stopped listening only to what people said and started watching what they did.

He noticed patterns.

Projects were framed as pilots—even when everyone knew they were intended to scale.
Emails were copied widely—not for transparency, but protection.
Questions about ownership were answered with org charts instead of names.

When something stalled, the explanation was rarely technical.

"It's still under review."
"We need one more sign-off."
"We're waiting on alignment."

Each delay was reasonable in isolation. Together, they formed a system that quietly discouraged momentum.

In CI (Continuous Improvement) terms, the work had variation—but no standard.

Month Three: Misdiagnosing Friction as Resistance

At the three-month mark, Alex began hearing a narrative he'd heard before in other organizations.

"The plants are pushing back."
"People are uncomfortable with the pace."
"We need stronger change management."

Alex didn't agree—but he also didn't argue.

Instead, he asked questions.

Where exactly is the resistance?
At what decision point does work stop?
What risk is being introduced—and to whom?

The answers were telling.

People weren't refusing to adopt technology. They were being asked to accept outcomes without clarity on responsibility, support, or escalation.

This wasn't fear.
It was rational caution inside an unclear system.

Month Four: Naming Friction Without Blame

Alex shifted his approach.

Instead of advocating for solutions, he began facilitating conversations—quietly and deliberately.

With Operations, he asked where technology made their work harder instead of easier.
With IT, he asked where governance protected the system and where it slowed learning.
With Finance, he asked how risk was evaluated—and who carried it after approval.

He didn't label these discussions as "transformation."
He framed them as friction identification.

That framing changed everything.

People spoke freely. Not emotionally—but precisely.

They didn't complain about tools.
They described uncertainty.

They described rework.
They described unclear handoffs.
They described decisions that felt personal instead of shared.

Friction surfaced because the system allowed it to.

What Friction Actually Is (CI Perspective)

In CI, friction is not resistance.
It is waste created by design.

Friction appears when:

- Ownership is unclear
- Decision rights are ambiguous
- Information arrives too late or in the wrong form
- Risk is unevenly distributed
- Systems demand workarounds to function

Friction is not emotional.
It is structural.

And like any form of waste, it compounds when ignored.

Month Five: Capturing Friction Systematically

Alex resisted the temptation to "fix" anything.

Instead, he documented.

He worked with OpX to create simple artifacts:

- Decision maps showing who actually approved what
- Intake paths for technology requests
- RACI models that reflected reality—not intent
- Playbooks that described how work should flow, end to end

These weren't heavy documents.
They were clarity tools.

For the first time, friction was visible.

Not as complaints.
As patterns.

Common Sources of Friction Identified

Across functions, the same gaps appeared repeatedly:

Ownership Friction
Multiple groups influenced outcomes, but no one owned them.

Decision Friction
Approvals existed, but escalation paths did not.

Information Friction
Data existed, but not where decisions were made.

Incentive Friction
Risk lived in one function; reward in another.

None of this reflected poor intent.
It reflected poor design.

Month Six: Collaboration Replaces Negotiation

Once friction was named, something shifted.

Conversations changed tone.

Instead of debating tools, leaders debated design.
Instead of defending roles, they clarified ownership.
Instead of slowing decisions, they improved decision quality.

Alex didn't "own" these changes.

OpX facilitated them.
IT refined governance.
Operations defined outcomes.
Finance clarified risk boundaries.

The system began to work with people instead of against them.

Why Transformation Feels Hard

Transformation feels hard when systems are unclear.

Not because people resist change—but because friction taxes attention, confidence, and trust.

What organizations often call fear is usually drag created by ambiguity.

Remove the ambiguity.
Friction falls.
Behavior follows.

The Core Insight

Friction in transformation is a design failure—not a people problem.

When leaders design systems that:

- Clarify ownership
- Align incentives
- Support decisions
- Respect the work

Transformation stops feeling heavy.

It starts feeling possible.

The Four Friction Zones

1. Ownership Friction

Core Question: Who owns the outcome?

- Accountability split across functions
- Success shared, failure isolated
- Responsibility implied, not explicit

Common mislabel:

"People aren't stepping up."

2. Decision Friction

Core Question: Who can decide — and how fast?

- Approval ≠ decision authority
- Escalation paths undefined
- Decisions revisited repeatedly

Common mislabel:

"We're not aligned yet."

3. Information Friction

Core Question: Is the right information available at the moment of decision?

- Data exists but lacks context
- Reporting lags reality
- Signals unclear or contradictory

Common mislabel:

"We need better visibility."

4. Incentive & Risk Friction

Core Question: Who absorbs risk if this fails?

- Risk centralized, reward distributed
- Leaders hesitate rationally
- Innovation framed as exposure

Common mislabel:

"Let's slow down."

Chapter 1 Summary

- Friction is not resistance.
- What appears as resistance is often rational behavior within the system.
- Systems—not people—create drag.
- Clarity reduces friction faster than persuasion.
- Adoption improves when design improves.

Questions for the Reader

- Where does work slow down without obvious resistance?
- Who owns outcomes when initiatives stall?
- Where is risk unclear or unevenly distributed?
- What behaviors is the system encouraging?
- Where does motivation get blamed for structural gaps?

These are not culture questions.
They are design questions.

Look for patterns in structure, not flaws in people.

What Comes Next

Recognizing friction is only the first step. Most organizations sense drag but cannot describe it precisely enough to address it.

In the next chapter, *Seeing Friction Clearly*, we examine how to make friction visible—and how shared language transforms vague frustration into actionable insight.

Chapter 2: Seeing Friction Clearly

Why Most Leaders miss Friction

Most leaders are not bad at recognizing problems.
They are bad at recognizing systemic problems early.

By the time friction is visible, it has already compounded. Meetings have multiplied. Trust has thinned. Workarounds have formed. Leaders begin reacting to symptoms instead of diagnosing causes.

The most common mistake leaders make at this stage is asking the wrong question.

They ask:
Why aren't people moving faster?

They should be asking:
Where is the system slowing them down?

Seeing friction clearly requires a shift—from managing behavior to examining design.

Friction Is Quiet at First

In its early stages, friction is subtle.

It does not announce itself as conflict.
It shows up as hesitation.

- A leader asks for more time
- A team asks for one more review
- A project is labeled a "pilot" indefinitely
- Decisions are deferred until "alignment" is achieved

None of this looks dysfunctional.

In fact, it often looks responsible.

That is why friction is so often misdiagnosed as prudence, culture, or readiness.

From Chapter 1: Alex Learns to Stop Listening for Resistance

By the fourth month at NorthRiver Foods, Alex stopped listening for complaints and started listening for patterns.

He noticed that when leaders hesitated, they used similar language—even across functions.

Operations spoke about exposure.
IT spoke about standards.
Finance spoke about justification.
OpX spoke about sustainability.

Different vocabularies.
The same underlying signal.

Each group was responding logically to unclear system design.

That realization changed how Alex listened—and how he led conversations.

The Shift: From Symptoms to Signals

In Operational Excellence, symptoms are not the problem.

They are indicators.

Late projects, stalled initiatives, endless pilots, and over-alignment are not failures of execution. They are signals of friction embedded in the system.

Seeing friction clearly means separating:

- What is happening
  from
- Why it is happening

This is where most transformation efforts fail.

The Friction Map as a Diagnostic Lens

To help leaders see friction without blame, Alex and the OpX team introduced a simple diagnostic structure: the Friction Map.

The Friction Map does not explain how work should flow.
It reveals where work actually slows—and why.

It forces four disciplined questions:

1. What work or decision are we trying to move?
2. Where does it slow or stop?
3. What design gap exists at that point?
4. How does that gap show up in behavior?

This structure does something powerful.

It keeps teams out of opinions and inside observation.

Friction Map
How friction shows up and what it impacts
Work / Decision
Where it Slows
The Design Gap
How it Shows Up
Friction Identified
EXAMPLES
New initiatives
Operational Execution
Investment Decisions
Acting on Data
Intake, prioritization
Approvals / Decisions
Post-approval Accountability
Reviewing / Interpretation
No clear owner end-to-end
Decision rights unclear
Risk assigned w/o authority
Data not available at decision gap
Endless pilots
"We need more alignment"
Delays - excess caution
Workarounds
Friction defined without blame - Turning Point

How to Read the Friction Map

The Friction Map is read left to right, not top to bottom.

You do not start with people.
You do not start with solutions.

You start with work.

This matters.

When leaders start with behavior, they almost always end with blame.
When they start with work, they end with design.

The Four Friction Zones

As Alex facilitated Friction Map discussions across NorthRiver, patterns emerged quickly. Friction clustered into four predictable zones.

These zones are not unique to technology or transformation.
They appear in any system under change.

Friction Zone 1: Ownership Friction

Defining question:
Who owns the outcome?

Ownership friction occurs when responsibility is implied but not explicit.

Common indicators:

- Multiple stakeholders influence decisions
- Accountability is shared for success but isolated for failure

- Leaders hesitate because consequences are unclear

What leaders often say:

"People need to step up."

What the system is actually saying:

"No one owns this end to end."

Ownership friction is the most dangerous form of friction because it creates silent avoidance. People stay involved—but avoid commitment.

Friction Zone 2: Decision Friction

Defining question:
Who can decide, and how fast?

Decision friction exists when approval paths exist, but decision rights do not.

Common indicators:

- Decisions are revisited repeatedly
- Escalation paths are undefined
- Authority exists on paper but not in practice

What leaders often say:

"We need better alignment."

What the system is actually saying:

"No one knows who can say yes."

Decision friction slows transformation more effectively than any technical limitation.

Friction Zone 3: Information Friction

Defining question:
Is the right information available at the moment of decision?

Information friction occurs when data exists but does not support action.

Common indicators:

- Data arrives after decisions are made
- Reports lack context
- Teams distrust the signal

What leaders often say:

"We need better visibility."

What the system is actually saying:

"Information is disconnected from decisions."

Information that cannot be acted upon is not visibility—it is noise.

Friction Zone 4: Incentive and Risk Friction

Defining question:
Who absorbs risk if this goes wrong?

Incentive friction exists when the system distributes reward but concentrates risk.

Common indicators:

- Leaders delay decisions despite clear benefits
- Innovation is framed as exposure
- Risk discussions dominate progress discussions

What leaders often say:

"Let's be careful."

What the system is actually saying:

"Risk is misaligned with authority."

This form of friction is often invisible to executives—and obvious to everyone else.

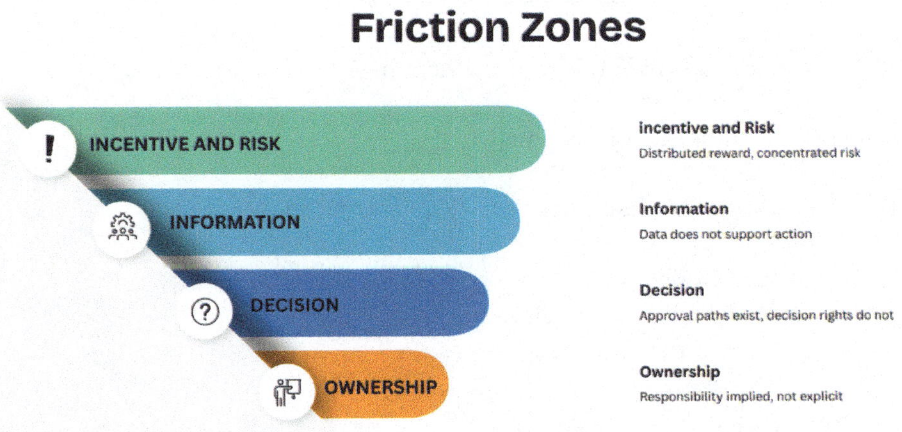

Why Leaders Misread Friction

Friction is often misread because leaders are rewarded for decisiveness, not diagnosis.

Acting quickly feels productive.
Slowing down to see friction feels like hesitation.

But in CI, we know this truth:

You cannot improve what you cannot see.

Seeing friction clearly is not slowing transformation.
It is preventing waste.

A Practical Exercise: Naming Friction Without Blame

Alex introduced a simple rule during Friction Map sessions:

No solutions until friction is named.

Teams were asked to describe:

- Where work slowed
- What decision was unclear
- What risk felt unowned

Not how to fix it.
Not who caused it.

Just where it existed.

The result was immediate.

Tension dropped.
Defensiveness faded.
Clarity emerged.

People don't resist diagnosis when it feels fair.

What Seeing Friction Enables

Once friction is visible:

- Conversations shift from persuasion to design
- Conflict turns into collaboration
- Change becomes less personal and more structural

This is the turning point.

Seeing friction clearly is what makes transformation possible.

Chapter 2 Summary

- Friction is a system signal, not a behavior flaw.
- Symptoms are data, not opposition.
- The Friction Map creates shared language.
- Visibility precedes redesign.
- Design—not motivation—drives adoption.

Questions for the Reader

- Where does work slow without obvious resistance?
- Which decisions are revisited most often?
- Where does risk feel personal instead of shared?
- Which friction zone appears most frequently?
- Where are symptoms being managed instead of mapped?

Do not answer these emotionally.
Answer them structurally.

Map patterns, not isolated events.

What Comes Next

Seeing friction clearly does not eliminate it. Once friction is visible, responsibility must shift from observation to ownership.

In the next chapter, *Ownership Is Not a Title*, we examine why formal authority does not guarantee accountability—and how unclear ownership sustains friction even when problems are well understood.

Chapter 3: Ownership Is Not a Title

Why Ownership Is the First Place Friction Hides

When transformation stalls, leaders often believe they have an ownership problem.

Someone isn't stepping up.
Someone isn't accountable.
Someone needs to lead this harder.

These assumptions feel reasonable. They are also usually wrong.

In most organizations, ownership does not fail because people avoid responsibility. It fails because the system never clearly assigns it.

Ownership is not a title.
It is a design choice.

And when it is poorly designed, friction becomes inevitable.

What Leaders Mean When They Say "Ownership"

In leadership discussions, ownership is often used as shorthand.

"We need someone to own this."
"Who owns the outcome?"
"This needs a single owner."

But when Alex listened closely to these conversations at NorthRiver Foods, he noticed something important.

Different leaders meant different things by the same word.

Operations believed ownership meant results.
IT believed ownership meant system integrity.
Finance believed ownership meant risk control.
OpX believed ownership meant sustainability.

Each perspective was valid.
None were complete.

The organization wasn't missing owners.
It was missing alignment on what ownership actually meant.

Month Seven: When Everyone Owns It, No One Does

As transformation efforts continued, Alex saw a recurring pattern.

Initiatives were staffed with steering committees.
Responsibility was shared across functions.
Updates were presented collectively.

From the outside, this looked collaborative.

In practice, it created a vacuum.

When milestones slipped, no one felt authorized to intervene decisively. When trade-offs emerged, decisions were deferred upward. When issues surfaced after launch, ownership quietly dissolved back into functional silos.

The system rewarded participation—not accountability.

This wasn't a people problem.
It was a design flaw.

Ownership as a System Property

In CI, ownership is not inferred from position.
It is explicitly designed.

True ownership has three inseparable elements:

1. Authority — the right to make decisions
2. Accountability — responsibility for outcomes
3. Support — access to resources and escalation

When any one of these is missing, ownership becomes symbolic instead of real.

Alex noticed that many transformation initiatives at NorthRiver assigned accountability without authority, or authority without support.

Both create friction.

The Most Common Ownership Design Failures

As Alex and the OpX team mapped friction across initiatives, the same ownership failures appeared repeatedly.

Failure Mode 1: Distributed Accountability

What it looks like:
Multiple leaders are "jointly accountable."

Why it creates friction:
Shared accountability often means diffused responsibility. Decisions slow because no one wants to overstep. Risks are managed cautiously because consequences are unclear.

What leaders say:
"This is cross-functional."

What the system creates:
Indecision.

Failure Mode 2: Ownership Without Authority

What it looks like:
A leader is named owner but lacks decision rights.

Why it creates friction:
The owner becomes a coordinator, not a leader. Progress depends on persuasion rather than action.

What leaders say:
"They own it, but IT still has to approve."

What the system creates:
Frustration and delay.

Failure Mode 3: Authority Without Accountability

What it looks like:
Decisions are made centrally, outcomes felt locally.

Why it creates friction:
Those affected by decisions disengage. Workarounds form. Trust erodes.

What leaders say:
"This was decided at a higher level."

What the system creates:
Quiet resistance.

Failure Mode 4: Ownership Ends at Approval

What it looks like:
Responsibility disappears after go-live.

Why it creates friction:
No one owns adoption, performance, or continuous improvement.

What leaders say:
"We implemented it."

What the system creates:
Stalled value.

How to Identify Ownership Friction in Your Organization

Ownership friction rarely announces itself directly. It hides behind good intentions, collaboration language, and well-meaning structures. Leaders often believe ownership exists because people are involved. In reality, involvement and ownership are not the same.

To identify ownership friction, leaders must stop asking who is working on this and start asking who is responsible for the outcome.

Observable Signals of Ownership Friction

Ownership friction is present when you consistently observe the following behaviors:

- Initiatives have many contributors but no single accountable owner
- Updates focus on activity, not outcomes
- Decisions are deferred "until alignment is reached"
- Issues surface late because no one feels authorized to escalate
- Ownership dissolves after approval or go-live

None of these indicate disengagement.
They indicate unclear system design.

Language That Signals Ownership Gaps

Listen carefully to how leaders describe work. Ownership friction has a vocabulary.

Common phrases include:

- "This is cross-functional"
- "We're all responsible"
- "It touches a lot of teams"
- "We need everyone aligned"

These statements are not wrong—but they often replace clarity with consensus.

When ownership is clear, leaders use names, not groups.

Diagnostic Questions Leaders Should Ask

To surface ownership friction, leaders should ask questions that force specificity:

- Who owns the outcome if this fails?
- Who decides when trade-offs arise?
- Who owns performance after implementation?
- Who is accountable when priorities conflict?

If answers differ depending on who you ask, ownership friction exists.

Where Ownership Friction Commonly Lives

Ownership friction tends to cluster in predictable places:

- Cross-functional initiatives
- Digital or technology-enabled change
- Work that spans IT, Ops, and Finance
- Initiatives labeled "strategic" or "transformational"

The more complex the work, the more explicit ownership must be.

What Not to Do When You See Ownership Friction

Leaders often respond to ownership friction in counterproductive ways:

- Adding steering committees
- Escalating decisions upward
- Asking for more collaboration
- Renaming roles without changing authority

These actions increase coordination—but rarely increase accountability.

Ownership friction is not solved by involvement.
It is solved by design.

How to Design Ownership That Reduces Friction

Once ownership friction is visible, the goal is not to assign blame or create new layers. The goal is to engineer clarity.

Effective ownership design focuses on outcomes, authority, and sustainment—not job titles.

Design Move 1: Assign Ownership to Outcomes, Not Activities

Ownership should be defined around results, not tasks.

Poor ownership sounds like:

- "IT owns the system"
- "OpX owns the process"
- "Operations owns execution"

Effective ownership sounds like:

- "This leader owns uptime improvement"
- "This role owns adoption and sustained use"
- "This function owns business value realization"

When ownership is tied to outcomes, decisions accelerate.

Design Move 2: Align Authority With Accountability

Accountability without authority creates frustration.
Authority without accountability creates detachment.

Ownership must include:

- Decision rights within defined boundaries
- Authority to resolve trade-offs
- Access to escalation when conflicts arise

If a leader must negotiate every decision upward or sideways, ownership is symbolic.

Design Move 3: Extend Ownership Beyond Approval

One of the most common ownership failures occurs at go-live.

Ownership often ends when:

- The project is approved
- The system is implemented
- The pilot concludes

True ownership includes:

- Adoption
- Performance
- Continuous improvement

If ownership ends at approval, value rarely materializes.

Design Move 4: Make Ownership Visible and Stable

Ownership should not live only in people's heads.

Effective systems:

- Document ownership explicitly
- Revisit it when conditions change
- Reference it when conflicts arise

This does not require bureaucracy.
It requires consistency.

When ownership is visible, friction decreases because escalation becomes predictable.

What Good Ownership Feels Like

When ownership is designed well:

- Decisions happen faster with less debate
- Leaders act without waiting for permission
- Issues surface earlier
- Accountability feels fair, not personal

Most importantly, people stop protecting themselves—and start protecting outcomes.

Early Warning Signs Ownership Is Regressing

Even well-designed ownership erodes over time.

Watch for:

- Increasing approvals
- Vague updates
- Reopened decisions
- Responsibility drifting back to groups

These are signals to revisit design—not to push harder.

Why Ownership Design Multiplies Everything Else

Ownership clarity reduces:

- Decision friction
- Information friction
- Incentive friction

Without ownership, no amount of data, governance, or tooling will sustain transformation.

Ownership is the load-bearing wall of system design.

How Alex Reframed the Ownership Conversation

Instead of asking who owns this, Alex began asking different questions:

- Who owns the decision?
- Who owns the outcome?
- Who owns the risk after launch?
- Who owns the learning when it doesn't work?

These questions changed the conversation.

They forced leaders to confront gaps between intent and design.

Ownership became something to engineer, not assume.

Designing Ownership That Reduces Friction

Through OpX workshops and charter sessions, Alex helped leaders define ownership explicitly.

Effective ownership design included:

- Clear decision rights documented—not implied
- Named owners for outcomes, not activities
- Explicit post-approval accountability
- Defined escalation paths when trade-offs arose

Most importantly, ownership was aligned to where work happened, not where hierarchy lived.

When ownership matched reality, friction decreased.

Why Ownership Feels Personal When It's Poorly Designed

One of the most damaging side effects of unclear ownership is emotional.

When systems are vague, people internalize risk.

Leaders hesitate because failure feels personal.
Teams avoid decisions because consequences are unclear.
Innovation slows because exposure outweighs reward.

This is often misread as fear.

In reality, it is rational behavior inside a poorly designed system.

Ownership Is the Foundation of Trust

Trust does not come from culture statements.
It comes from predictability.

When people know:

- Who decides
- Who supports
- Who owns the outcome

They move faster.

Ownership clarity does not eliminate disagreement—but it removes hesitation.

Chapter 3 Summary

- Ownership is a system design choice.
- Titles do not create accountability.
- Authority, accountability, and support must align.
- Involvement does not equal ownership.
- Poor ownership design creates silent friction.
- Clear ownership reduces drag across the system.

Questions for the Reader

- Where is ownership implied instead of explicit?
- Who carries accountability without authority?
- Where does ownership dissolve after launch?
- Where is accountability shared but authority unclear?
- Who owns outcomes after approval?
- Which leaders carry risk without control?

These are not leadership style questions.
They are system design questions.

Do not fix tools.
Fix ownership design.

What Comes Next

Ownership clarity is necessary—but not sufficient. Even when roles are defined, transformation slows when the decisions themselves are poorly structured.

In the next chapter, When Approval Replaces Decision, we examine Decision Friction—and why approval-heavy systems create the appearance of control while quietly undermining progress.

Chapter 4: When Approval Replaces Decision

Why Organizations Confuse Approval With Progress

In most organizations, approvals feel safe.

They provide structure.
They signal diligence.
They create the appearance of control.

When transformation slows, leaders often respond by adding more approvals, more reviews, and more checkpoints. Each addition feels responsible. Each one reduces individual exposure.

Collectively, they create friction.

Approval-heavy systems are rarely designed intentionally. They accumulate quietly over time—each one added in response to a past failure, audit finding, or missed expectation.

What begins as risk management eventually becomes decision paralysis.

The Hidden Cost of Approval Culture

Alex encountered this early at NorthRiver Foods.

On paper, the organization had a clear approval structure. Requests flowed upward. Reviews were documented. Sign-offs were tracked. Nothing moved without visibility.

Yet initiatives stalled at predictable points:

- After initial intake
- During cross-functional review
- Before final authorization

Each delay was justified.
None were owned.

Alex noticed something subtle: approvals existed everywhere, but decisions existed nowhere.

Approval Is Not a Decision

This distinction is critical—and often misunderstood.

An approval confirms compliance.
A decision commits to an outcome.

Approvals answer questions like:

- Does this meet standards?
- Has risk been reviewed?
- Is funding available?

Decisions answer different questions:

- Are we doing this?
- What trade-offs are we accepting?
- Who owns the outcome if it fails?

When organizations substitute approvals for decisions, they create the illusion of progress while avoiding accountability.

Month Eight: When No One Can Say Yes

By the eighth month, Alex could predict where initiatives would slow.

Not because of technology.
Not because of resistance.

But because no one was clearly empowered to decide.

Leaders waited for consensus.
Consensus waited for approvals.
Approvals waited for alignment.

Alignment waited for someone else to move first.

This loop didn't create conflict.
It created inertia.

Why Approval-Based Systems Feel So Comfortable

Approval systems persist because they solve an emotional problem for leaders.

They reduce personal risk.

When outcomes are uncertain, approvals distribute responsibility.
When something goes wrong, leaders can point to the process.

"I followed the process."
"It was approved."
"We had alignment."

The process becomes a shield.

But shields also block movement.

Decision Friction Defined

Decision friction occurs when:

- Authority exists, but is not exercised
- Approval paths are clear, but decision rights are not
- Escalation exists only upward
- Decisions are revisited after being "approved"

Decision friction does not look chaotic.
It looks disciplined.

That is why it is so dangerous.

How to Identify Decision Friction in Your Organization

Decision friction is often mistaken for diligence. From the outside, it looks like caution, alignment, or governance at work. From the inside, it feels like momentum draining away.

Decision friction exists when decisions technically move—but never fully land.

Leaders leave meetings believing something was decided, only to revisit the same topic weeks later under slightly different conditions.

Observable Signals of Decision Friction

Decision friction is present when you see patterns such as:

- Decisions are revisited after being "approved"
- Meetings focus on alignment rather than commitment
- Initiatives stall at predictable approval points
- Escalation feels unsafe or unclear
- Leaders hesitate to act without broad consensus

None of these indicate poor leadership.

They indicate unclear decision design.

Language That Signals Decision Friction

Decision friction has a recognizable vocabulary. Listen for phrases like:

- "We're close, but not quite there yet"
- "Let's socialize this a bit more"
- "We need one more review"
- "This is bigger than we thought"

These statements often appear when authority exists—but is not usable.

Diagnostic Questions Leaders Should Ask

To surface decision friction, leaders should ask questions that clarify authority and consequence:

- Who can make this decision today?
- What happens if we decide incorrectly?
- Who supports the decision-maker after commitment?
- At what point does this escalate—and to whom?

If answers rely on hierarchy rather than design, friction is present.

Where Decision Friction Commonly Lives

Decision friction clusters in places where:

- Risk is high
- Visibility is broad
- Accountability is unclear

Common hotspots include:

- Capital investments
- Digital or technology-enabled initiatives
- Cross-functional prioritization
- Architecture or standard-setting decisions

The more consequential the decision, the more explicit the design must be.

What Not to Do When You See Decision Friction

Leaders often respond to decision friction by adding structure—but not clarity.

Common missteps include:

- Adding approval layers
- Creating additional forums
- Seeking more alignment
- Deferring decisions upward

These actions slow work without reducing risk.

Decision friction is not solved by more control.
It is solved by clear decision rights.

How to Design Decisions That Reduce Friction

Well-designed decisions feel lighter—not because they are easier, but because expectations are clear.

Designing decisions does not mean removing governance.
It means ensuring governance supports commitment, not avoidance.

Design Move 1: Explicitly Name the Decision Owner

Every recurring decision should have a single, named owner.

This does not mean others are excluded.
It means one person is accountable for making the call.

Decision ownership should answer:

- Who decides?
- Who provides input?
- Who must be informed?

When decision ownership is explicit, alignment happens faster—and quieter.

Design Move 2: Separate Reversible From Irreversible Decisions

Not all decisions carry the same risk.

Treating all decisions as irreversible forces unnecessary caution.

Effective design:

- Identifies which decisions can be reversed safely
- Encourages speed where learning is possible
- Reserves heavy governance for truly irreversible choices

This distinction alone removes enormous friction.

Design Move 3: Define Escalation Before It's Needed

Escalation is not failure.
Undefined escalation is.

Decision design should specify:

- When escalation is appropriate
- Who escalation goes to
- What support follows escalation

When escalation paths are clear, leaders act sooner.

Design Move 4: Limit Approvals to Risk Boundaries

Approvals should protect the organization—not replace decisions.

Effective systems:

- Use approvals to guard true risk boundaries
- Avoid approvals that duplicate decision authority
- Remove approvals that exist "just in case"

When approvals are purposeful, decision confidence increases.

What Good Decision Design Feels Like

When decision friction is reduced:

- Meetings end with commitments
- Decisions are not reopened casually
- Leaders act without waiting for consensus
- Trade-offs are explicit and accepted

Disagreement still exists.
Indecision does not.

Early Warning Signs Decision Friction Is Returning

Decision friction tends to creep back quietly.

Watch for:

- Increasing pre-meetings
- Repeated requests for alignment
- Decisions reframed as recommendations
- Authority drifting upward

These are signals to revisit decision design—not to push urgency.

Why Decision Design Accelerates Trust

Clear decision rights reduce:

- Political behavior
- Defensive communication
- Over-alignment

People trust systems where decisions are predictable—even when outcomes are uncertain.

Trust grows when leaders know:

- Who decides
- How decisions are supported
- What happens if things go wrong

How Decision Friction Shows Up

Across organizations, decision friction produces the same observable behaviors:

- Projects labeled as "pilots" indefinitely
- Decisions reopened after new stakeholders engage
- Meetings focused on alignment instead of action
- Leaders avoiding explicit trade-offs

These behaviors are often misinterpreted as collaboration.

They are actually avoidance mechanisms created by system design.

What Leaders Often Say—and What the System Is Really Saying

Leader language:
"We need more alignment."

System signal:
No one is authorized to decide.

Leader language:
"Let's slow down and be careful."

System signal:
Risk is unclear or unevenly distributed.

Leader language:
"We need to socialize this more."

System signal:
Escalation paths are undefined.

Seeing these patterns for what they are allows leaders to respond differently.

Why Decision Friction Feels Like Culture—but Isn't

Decision friction is often blamed on culture because its effects are behavioral.

People hesitate.
Leaders defer.
Teams wait.

But culture does not create decision rules.

Systems do.

When decision authority is unclear, even strong cultures slow down. When escalation paths are vague, even decisive leaders hesitate.

Culture amplifies systems—it does not replace them.

Alex Changes the Question

Instead of asking:
"Why aren't we deciding faster?"

Alex began asking:

- Who is allowed to decide this?
- What decision is actually being made?
- What happens if this decision is wrong?
- Who supports the decision-maker afterward?

These questions shifted conversations immediately.

They surfaced gaps no one had named before.

Designing Decisions Instead of Managing Approvals

As OpX facilitated decision design discussions, several principles emerged:

- Every recurring decision needs a named decision owner
- Decision rights must be explicit—not inferred
- Escalation paths must be defined before they are needed
- Approval steps should support decisions—not replace them

This required letting go of some perceived safety.

But it increased speed, clarity, and trust.

The Cost of Reversible vs. Irreversible Decisions

One of the most powerful reframes Alex introduced was separating decisions into two categories:

- Reversible decisions — safe to make quickly
- Irreversible decisions — require deliberation

Most organizations treat all decisions as irreversible.

This is a design flaw.

When reversible decisions require approval-heavy processes, friction multiplies and learning slows.

What Happens When Decisions Are Designed Well

As decision authority clarified at NorthRiver:

- Meetings shortened
- Rework decreased
- Leaders stopped revisiting settled decisions
- Trust increased across functions

Not because people changed.

Because the system did.

Decision Clarity Reduces Fear Without Addressing Fear

Interestingly, as decisions became clearer, many leaders noticed something unexpected.

Tension dropped.

Not because people were reassured—but because uncertainty was removed.

This reinforced Alex's core insight:

What looks like fear is often just friction created by poor decision design.

Chapter 4 Summary

- Approvals do not equal decisions.
- Approval-heavy systems dilute accountability.
- Decision friction often presents as discipline.
- Excessive approval creates the illusion of control.
- Clear decision rights accelerate trust and execution.

Questions for the Reader

- Which decisions are revisited most often?
- Where do approvals substitute for commitment?
- Who is empowered to say yes—and supported after?
- Which decisions could safely be reversible?
- Where does risk escalate because decisions are delayed?

These are not leadership courage questions.
They are system design questions.

Examine the structure of the decision, not the personality of the leader.

What Comes Next

Even with clear ownership and defined decision rights, transformation can stall when information fails to reach the point of action—or arrives without context.

In the next chapter, When Visibility Isn't Clarity, we examine Information Friction—and why more dashboards, reports, and metrics do not automatically improve decisions.

Chapter 5: When Visibility Isn't Clarity

Why More Data Often Makes Things Worse

Most transformation efforts promise visibility.

Dashboards.
Reports.
KPIs.
Real-time data.

Leaders approve investments with the expectation that once information is visible, decisions will improve and performance will follow.

But in many organizations, the opposite happens.

As visibility increases, confidence decreases.
Decisions slow.
Debates intensify.
Trust erodes.

This is not a failure of analytics.
It is a failure of information design.

Visibility Is Not the Same as Clarity

Visibility answers the question:
Can I see it?

Clarity answers a different question:
Can I act on it?

Organizations often confuse the two.

They invest heavily in making data available without considering:

- Who needs the information
- When they need it
- What decision it is meant to support
- What action should follow

When information is disconnected from decisions, it creates friction instead of flow.

Month Nine: Drowning in Data

By the ninth month at NorthRiver Foods, Alex noticed a paradox.

Leaders had more data than ever before—and less confidence using it.

Dashboards showed trends, but not context.
Reports were accurate, but late.
Metrics were debated instead of trusted.

Meetings shifted from "What should we do?"
to
"Which number is right?"

Information was visible.
Decision-making was not clearer.

Information Friction Defined

Information friction occurs when:

- Data exists but does not support action
- Signals arrive too late to influence decisions
- Metrics are disconnected from operational reality

- Interpretation varies by function

Information friction does not look like ignorance.
It looks like over-analysis.

How Information Friction Shows Up

Across organizations, information friction produces predictable behaviors:

- Leaders ask for "one more report"
- Teams build shadow spreadsheets
- Metrics are debated rather than acted upon
- Data is trusted selectively

These behaviors are often labeled as alignment issues.

They are actually symptoms of poor information design.

—

What Leaders Say—and What the System Is Really Saying

Leader language:
"We need better visibility."

System signal:
Information is not arriving at the decision point.

Leader language:
"We don't trust the data."

System signal:
Data lacks context or ownership.

Leader language:
"The dashboard doesn't tell the whole story."

System signal:
Metrics are disconnected from the work.

Seeing this distinction matters.

How to Identify Information Friction in Your Organization

Information friction is one of the most deceptive forms of friction because it often looks like progress. Dashboards are built. Reports are automated. Data pipelines exist.

And yet—decisions still slow.

To identify information friction, leaders must stop asking what data do we have? and start asking what decisions does this data actually support?

Observable Signals of Information Friction

Information friction is present when you consistently see:

- Leaders debate numbers instead of deciding actions

- Reports arrive after decisions are already made
- Teams maintain shadow spreadsheets "to be safe"
- Metrics are reviewed but rarely acted upon
- Different functions tell different stories with the same data

None of these indicate poor analytics capability.
They indicate misaligned information design.

Language That Signals Information Friction

Information friction has a recognizable vocabulary.

Listen for phrases such as:

- "The data doesn't tell the whole story"
- "We don't trust this number yet"
- "We need more context"
- "Let's wait for the next report"

These statements usually signal that information is disconnected from decision-making—not that it is inaccurate.

Diagnostic Questions Leaders Should Ask

To surface information friction, leaders should ask questions that force alignment between data and decisions:

- What decision is this metric meant to support?
- Who is expected to act when this number changes?
- How quickly does this information need to arrive?
- What action should follow if this signal worsens or improves?

If these questions cannot be answered clearly, information friction exists.

Where Information Friction Commonly Lives

Information friction tends to concentrate in areas where:

- Data spans multiple systems or functions
- Metrics are owned centrally but used locally
- Performance discussions outpace operational reality
- Visibility is prioritized over action

Common examples include:

- Enterprise dashboards
- Weekly performance reports
- Cross-functional KPIs
- Digital transformation metrics

The broader the audience, the more intentional the design must be.

What Not to Do When You See Information Friction

Leaders often respond to information friction by increasing volume.

Common missteps include:

- Adding more metrics
- Increasing reporting frequency
- Launching new dashboards
- Requesting deeper analysis

These actions increase cognitive load without increasing clarity.

Information friction is not solved by more data.
It is solved by better alignment between data and decisions.

How to Design Information That Reduces Friction

Designing information for clarity requires discipline. The goal is not to inform everyone—it is to enable the right decisions at the right time.

Design Move 1: Start With the Decision, Not the Metric

Every piece of information should exist to support a specific decision.

Effective information design begins by asking:

- What decision is being made?
- Who makes it?
- What signal matters most?

Metrics that do not serve a decision should be questioned—or removed.

Design Move 2: Assign Ownership for Interpretation, Not Just Reporting

Data ownership is not the same as interpretation ownership.

Effective systems clearly define:

- Who ensures the data is accurate
- Who explains what it means

- Who acts when it changes

When interpretation is unclear, hesitation follows—even when data is trusted.

Design Move 3: Align Information Timing With Decision Timing

Late information creates the illusion of visibility while undermining action.

Information should arrive:

- Before decisions—not after
- At the point of use—not in hindsight
- With enough lead time to respond

If decisions must be made weekly, monthly data creates friction.

Design Move 4: Define Expected Responses to Signals

Metrics without defined responses invite debate.

Effective information design specifies:

- What action is expected when a signal crosses a threshold
- Who initiates that action
- How escalation occurs if results persist

This turns data into guidance—not discussion.

What Good Information Design Feels Like

When information friction is reduced:

- Leaders stop asking for “one more report”
- Meetings focus on decisions, not explanations
- Data confirms reality instead of contradicting it
- Action follows signal quickly and predictably

Information becomes an enabler—not a distraction.

Early Warning Signs Information Friction Is Returning

Information friction often creeps back quietly.

Watch for:

- Growing dashboards
- Increasing explanation time in meetings
- Repeated disputes over metrics
- Requests for manual validation

These are signals to revisit information design—not to add analytics.

Why Information Design Multiplies Decision Quality

Well-designed information:

- Reduces decision friction
- Builds trust without persuasion

- Decreases over-alignment
- Accelerates learning

Poorly designed information does the opposite—no matter how accurate it is.

Why More Metrics Rarely Solve the Problem

When leaders sense uncertainty, the instinctive response is to measure more.

More KPIs.
More dashboards.
More reporting cadence.

This often increases friction.

Each new metric adds cognitive load. Each new report introduces another interpretation. Each new dashboard becomes another artifact to reconcile.

The problem is not volume.
It is relevance.

-

Information Is a Form of Standard Work

In CI, information is not neutral.

It either:

- Enables correct action
  or
- Creates confusion and waste

Alex began reframing information as standard work for decisions.

That simple shift changed conversations.

Instead of asking "What data do we have?" teams asked:

- What decision is this data meant to support?
- What action should follow if the signal changes?
- Who owns the interpretation?

Information became purposeful.

Where Information Friction Is Born

As Alex and the OpX team mapped information flow, four design gaps appeared repeatedly:

Timing Gaps
Data arrived after decisions were already made.

Context Gaps
Metrics lacked operational explanation.

Ownership Gaps
No one owned interpretation, only collection.

Action Gaps
No clear response was defined when metrics moved.

Each gap added friction.

Together, they stalled progress.

Why Information Friction Feels Like a Trust Problem

Information friction often gets labeled as a trust issue.

"We don't trust the numbers."
"They don't reflect reality."
"This doesn't match what we're seeing."

But trust is not built by accuracy alone.

Trust is built when information:

- Matches lived experience
- Supports timely decisions
- Leads to consistent action

When those conditions aren't met, skepticism is rational.

Alex Stops Asking for Better Data

Instead of pushing for better dashboards, Alex asked different questions:

- Where is this decision being made?
- What information is needed at that moment?
- What happens when the signal changes?
- Who is accountable for acting?

These questions surfaced more value than any report ever had.

Designing Information for Decisions

As OpX facilitated information redesign, several principles emerged:

- Every metric must have a decision owner
- Every decision must have a defined signal
- Every signal must have an expected response
- Information should reduce debate, not create it

When these principles were applied, something surprising happened.

Leaders stopped asking for more data.

They started acting.

The Cost of Ignoring Information Friction

When information friction persists:

- Decisions drift upward
- Frontline autonomy erodes
- Confidence declines
- Transformation slows quietly

None of this appears dramatic.

It just feels heavy.

Chapter 5 Summary

- Visibility does not equal clarity.
- Information must serve decisions.
- Metrics without context create friction.
- Metrics without action create drag.
- Information design is leadership work.
- Information is standard work for leaders.

Questions for the Reader

If data feels heavy, ask:

- Which metrics drive decisions—and which don't?
- Where does information arrive too late to matter?
- Who owns interpretation?
- What action should follow each signal?

These are not analytics questions.
They are system design questions.

Do not ask for better dashboards.
Redesign information flow.

What Comes Next

Even with clear ownership, defined decision rights, and well-designed information flow, transformation can still stall when risk and incentives are misaligned.

In the next chapter, When Risk Feels Personal, we examine Incentive and Risk Friction—and why rational leaders hesitate even when everything else appears clear.

Chapter 6: When Risk Feels Personal

Why Rational Leaders Hesitate

By the time Alex reached his tenth month at NorthRiver Foods, something became clear.

Even when ownership was defined, decisions clarified, and information improved, certain initiatives still moved slowly.

Not stalled.
Not blocked.
Just… cautious.

Leaders asked careful questions.
They requested additional validation.
They delayed commitments without outright refusal.

This behavior was often described as risk aversion.

Alex saw it differently.

It was risk misalignment.

Risk Is Not the Same as Uncertainty

In transformation discussions, risk and uncertainty are often used interchangeably.

They are not the same.

Uncertainty comes from lack of information.
Risk comes from consequences.

Information friction creates uncertainty.
Incentive friction creates hesitation—even when information is clear.

This distinction matters.

Incentive and Risk Friction Defined

Incentive and risk friction occurs when:

- Leaders carry risk without commensurate authority
- Outcomes are shared, but consequences are individual
- Failure is visible; success is normalized
- Career exposure outweighs operational benefit

This friction is quiet.

People do not resist.
They slow down.

How to Identify Incentive and Risk Friction in Your Organization

Incentive and risk friction is the hardest friction to see because it is rarely spoken aloud. Leaders rarely say they are hesitant because of personal exposure. Instead, hesitation shows up as caution, delay, or requests for more certainty.

To identify incentive and risk friction, leaders must stop asking why people are slow and start asking who absorbs the downside when things go wrong.

Observable Signals of Incentive and Risk Friction

Incentive and risk friction is present when you observe patterns such as:

- Leaders delaying commitment despite strong business cases
- Preference for pilots that never scale
- Requests for executive sponsorship late in the process
- Decisions pushed upward even when authority exists
- Increased caution around highly visible initiatives

None of these indicate fear or lack of confidence.
They indicate misaligned risk ownership.

Language That Signals Incentive and Risk Friction

This form of friction has a subtle but consistent vocabulary.

Listen for phrases like:

- "This could have broader implications"
- "Let's make sure leadership is aligned"
- "We don't want to get ahead of ourselves"
- "This is a big bet"

These statements often appear when leaders believe success will be normalized—but failure will be remembered.

Diagnostic Questions Leaders Should Ask

To surface incentive and risk friction, leaders should ask questions that clarify exposure and support:

- Who is personally exposed if this fails?
- How visible will failure be compared to success?
- What support exists after approval?
- How is learning treated when outcomes fall short?

If these questions feel uncomfortable, incentive friction likely exists.

Where Incentive and Risk Friction Commonly Lives

Incentive and risk friction tends to concentrate in areas where:

- Decisions are highly visible
- Outcomes affect career perception
- Benefits accrue later than consequences
- Risk crosses functional boundaries

Common examples include:

- Digital transformation initiatives
- Capital investments
- Cross-site standardization efforts
- Enterprise-wide technology changes

The higher the visibility, the more intentional risk design must be.

What Not to Do When You See Incentive and Risk Friction

Leaders often respond to hesitation by encouraging courage or urgency.

Common missteps include:

- Pushing for faster decisions
- Framing hesitation as resistance
- Elevating tone instead of redesigning support
- Overusing pilots as safety nets

These approaches increase pressure without reducing exposure.

Incentive friction is not solved by motivation.
It is solved by system-level risk design.

How to Design Incentives and Risk That Reduce Friction

Reducing incentive friction does not mean eliminating risk. It means distributing it fairly and intentionally.

Well-designed systems make it clear that leaders will not be left alone with downside.

Design Move 1: Make Post-Decision Support Explicit

One of the most powerful friction-reducing moves is clarifying what happens after a decision is made.

Effective systems explicitly define:

- Who supports the decision-maker
- How issues are escalated
- What resources are committed if challenges arise

When leaders know they won't be isolated, they act sooner.

Design Move 2: Normalize Learning, Not Just Success

When failure is punished and success is expected, hesitation is rational.

Effective organizations:

- Treat learning as an outcome
- Review decisions, not just results
- Separate intent from execution

This does not reduce accountability.
It reduces fear disguised as caution.

Design Move 3: Align Risk With Authority

Risk should sit where authority exists.

In poorly designed systems:

- Leaders carry risk without control
- Decisions feel career-defining
- Escalation becomes avoidance

Effective design ensures:

- Authority matches exposure
- Decision-makers can influence outcomes
- Risk is shared across the system

When risk alignment improves, speed follows naturally.

Design Move 4: Define the Path From Pilot to Scale

Pilots are often used to manage risk—but without clear rules, they become friction.

Effective systems define:

- What success looks like
- When scaling occurs
- Who decides to scale
- What support follows scale

This turns pilots into learning mechanisms—not delay tactics.

What Good Risk Design Feels Like

When incentive friction is reduced:

- Leaders commit earlier
- Pilots transition to scale
- Decisions feel supported
- Confidence replaces caution

Risk does not disappear.
But it no longer feels personal.

Early Warning Signs Incentive Friction Is Returning

Incentive and risk friction tends to creep back during pressure.

Watch for:

- Increasing reliance on pilots
- Decisions reframed as recommendations
- Requests for executive air cover
- Risk discussions replacing outcome discussions

These are signals to revisit risk design, not to push urgency.

Why Incentive Design Unlocks Everything Else

Even with:

- Clear ownership
- Strong decision rights
- Good information

Transformation will stall if leaders believe they carry disproportionate downside.

Risk alignment is the final unlock.

Month Ten: The Moment Everything Looked Ready—and Still Didn't Move

Alex watched an initiative reach a familiar point.

The business case was solid.
Operations supported it.
IT approved the architecture.
Data was available.

And yet, progress paused.

When Alex spoke with leaders individually, the hesitation wasn't technical.

It was personal.

"What happens if this fails?"
"Who owns the downside?"
"How visible will this be?"

No one said they were afraid.

They didn't have to.

Why This Friction Is So Often Invisible

Incentive friction rarely shows up in dashboards or project plans.

It shows up in:

- Delayed commitments
- Requests for additional validation
- Preference for pilots over scale
- Deferring decisions upward

Because this behavior is professional and polite, it often goes unchallenged.

But its impact is significant.

What Leaders Say—and What the System Is Really Saying

Leader language:
"Let's be careful."

System signal:
Risk and authority are misaligned.

Leader language:
"We need executive support."

System signal:
The decision carries personal exposure.

Leader language:
"This could have broader implications."

System signal:
Failure will be remembered longer than success.

Understanding this translation is essential.

Why Incentives Matter More Than Intent

Most leaders want transformation to succeed.

But intent does not override incentives.

When systems:

- Penalize failure
- Reward caution
- Publicize missteps
- Normalize wins

They teach people to slow down.

This is not a cultural flaw.
It is rational adaptation.

Alex Stops Pushing—and Starts Asking About Risk

Instead of encouraging urgency, Alex changed the conversation.

He asked:

- Who is exposed if this fails?
- What support exists after approval?
- How visible is this decision compared to its upside?
- What learning is protected?

These questions surfaced truths that had gone unspoken.

Common Sources of Incentive and Risk Friction

As OpX mapped friction, four patterns appeared consistently:

Asymmetric Risk
Failure is owned by individuals; success is shared broadly.

Delayed Reward
Benefits accrue later; consequences are immediate.

Visibility Bias
Mistakes are highlighted; improvements are normalized.

Career Exposure
Risk impacts perception, not just performance.

Each reinforces hesitation.

Why Pilots Feel Safer Than Progress

Pilots are often framed as learning tools.

In reality, they are sometimes risk shields.

Pilots:

- Limit exposure
- Avoid commitment
- Reduce visibility
- Defer accountability

This does not mean pilots are bad.

It means they are often overused to compensate for incentive friction.

Designing for Shared Risk

As Alex worked with leadership, a key principle emerged:

Risk must be designed—not absorbed.

This meant:

- Explicit post-approval support
- Clear escalation paths
- Shared accountability for outcomes
- Normalizing learning, not just success

When leaders knew they wouldn't be left alone with failure, behavior changed.

Why Removing Incentive Friction Unlocks Speed

Once incentive friction was addressed:

- Decisions accelerated
- Pilots transitioned to scale
- Leaders acted with confidence
- Trust increased across functions

Not because risk disappeared.

But because it was shared.

The Deeper Truth

What looks like fear is often just misaligned incentives.

When systems make risk personal and reward distant, hesitation is logical.

Transformation stalls not because leaders lack courage—but because systems punish it.

Chapter 6 Summary

- Risk misalignment creates hesitation.
- Incentives shape behavior more than intent.
- Motivation does not overcome exposure.
- Pilots often mask deeper incentive friction.
- Incentive friction is rational behavior.
- Shared risk accelerates transformation.

Questions for the Reader

- Who carries the downside when initiatives fail?
- Where is risk personal but reward collective?
- Which decisions feel career-defining?
- How visible is failure compared to success?
- What support follows commitment?
- How is learning treated when outcomes fall short?

These are not mindset questions.
They are system design questions.

Do not demand courage.
Redesign risk and incentives.

What Comes Next

At this point, friction is fully visible—ownership, decisions, information, and risk. Yet awareness alone does not create change.

In the next chapter, Why Seeing Friction Isn't Enough, we examine why diagnosis must be followed by deliberate system redesign—and why many transformations stall at insight.

Chapter 7: Why Seeing Friction Isn't Enough

Awareness Does Not Equal Change

By the time leaders can clearly name friction, something important has already happened.

Language has improved.
Conversations are more honest.
Blame has softened.

And yet—progress still stalls.

This is the moment where many transformation efforts lose momentum. Leaders feel they finally understand the problem, but the system continues to behave the same way.

Seeing friction is necessary.
It is not sufficient.

The False Finish Line

Organizations often mistake diagnosis for resolution.

Once friction is named, leaders expect behavior to change naturally. They assume clarity will translate into action. When it doesn't, frustration returns—sometimes sharper than before.

"We've talked about this."
"Everyone agrees."
"So why is nothing changing?"

The answer is simple, and uncomfortable:

Systems do not change because they are understood.
They change because they are redesigned.

Month Eleven: The Plateau

At NorthRiver Foods, Alex noticed a shift after several months of friction mapping.

Leaders spoke openly about ownership gaps.
Decision bottlenecks were acknowledged.
Information issues were visible.
Risk misalignment was named.

Yet initiatives still followed old paths.

Approvals accumulated.
Pilots lingered.
Escalations moved upward.

The organization had reached a plateau—not of understanding, but of design inertia.

Why Systems Resist Change Even When Leaders Don't

Systems are not neutral.

They encode:

- Past failures
- Historical compromises
- Unspoken rules
- Risk-avoidance mechanisms

These elements persist long after the original reasons are forgotten.

Even when leaders want change, systems continue to reward old behavior.

This is not sabotage.
It is structural memory.

The Comfort of Familiar Friction

One of the most overlooked dynamics in transformation is this:

Friction, once familiar, feels safer than redesign.

Unclear ownership is frustrating—but predictable.
Approval-heavy processes are slow—but known.
Ambiguous risk is uncomfortable—but survivable.

Redesign introduces uncertainty.

People may complain about friction, but they understand how to navigate it. New systems require relearning—not just new behaviors, but new expectations.

Why Intent Fails Without Design

Good intentions are not enough to overcome system design.

Leaders may agree to:

- Move faster
- Empower teams
- Reduce approvals
- Share risk

But unless those intentions are embedded into structures, behavior will revert under pressure.

In moments of stress, people default to the system—not the agreement.

Alex Encounters the Limits of Alignment

Alex saw this firsthand.

After productive workshops and aligned leadership conversations, teams returned to old habits. Not because they disagreed—but because the system still demanded it.

Approvals were still required.
Escalation paths were unchanged.
Risk remained unevenly distributed.

The organization was aligned in principle—and constrained in practice.

The Difference Between Agreement and Enablement

Agreement is verbal.
Enablement is structural.

Organizations are often rich in alignment and poor in enablement.

Enablement requires:

- Redefined roles
- Explicit decision rights
- Updated governance

- New operating rhythms

Without these, friction remains—even when everyone wants it gone.

Why Leaders Underestimate Design Work

Design work feels slow.

It requires:

- Letting go of legacy controls
- Making trade-offs explicit
- Reducing optionality
- Accepting new risks

Compared to launching initiatives or approving tools, design work feels abstract and uncomfortable.

But it is the only work that actually removes friction.

The Trap of Incremental Adjustment

Many organizations attempt to reduce friction incrementally.

They:

- Add clarity without removing approvals
- Introduce dashboards without changing decisions
- Name owners without granting authority
- Encourage risk-taking without changing incentives

These adjustments reduce surface tension—but leave the core design intact.

Friction persists.

Design Is an Act of Leadership

Redesigning systems is not administrative work.

It is leadership work.

It requires leaders to:

- Decide what matters most
- Accept constraints intentionally
- Make authority visible
- Share risk explicitly

This work cannot be delegated entirely.
It must be sponsored and protected.

The Moment the Book Turns

Up to this point, the book has focused on seeing.

Seeing friction.
Seeing patterns.
Seeing systems clearly.

From here forward, the focus shifts.

The next chapters are about designing friction out—intentionally, repeatably, and sustainably.

How Leaders Should Act Once Friction Is Visible

Seeing friction clearly creates a powerful—and dangerous—moment for leaders.

Once patterns are named, the instinct is to fix them immediately. Ownership feels unclear, so leaders assign it. Decisions feel slow, so approvals are removed. Risk feels personal, so escalation is mandated.

This impulse is understandable.

It is also where many transformation efforts fail.

The goal at this stage is not speed.
The goal is intentional movement without overcorrection.

The Most Common Mistake After Diagnosis

When friction becomes visible, leaders often try to remove it all at once.

They:

- Redesign governance broadly
- Rewrite decision rules
- Launch new operating models
- Communicate urgency

These actions feel decisive.

They also introduce instability.

Friction accumulated over time.
It cannot be removed responsibly in a single motion.

Why Overcorrection Creates New Friction

Systems are interconnected.

Ownership, decisions, information, and incentives reinforce one another. When one element changes abruptly, others strain to compensate.

For example:

- Removing approvals without redefining decision authority increases anxiety
- Naming owners without clarifying support increases risk exposure
- Improving information flow without decision clarity increases debate

The result is not flow.
It is a different kind of friction.

The Leader's Real Role at This Stage

Once friction is visible, leadership shifts.

The leader's role is no longer to diagnose.
It is to sequence design intentionally.

This requires restraint.

Effective leaders resist the urge to fix everything and instead focus on:

- What must change first
- What must remain stable
- What people need to feel safe moving forward

This is design leadership, not execution leadership.

What Should Change First—and What Shouldn't

Not all friction deserves immediate attention.

At this stage:

- Address friction that blocks decisions, not discomfort
- Prioritize clarity over optimization
- Change structures before changing behavior

What should not change yet:

- Reporting relationships
- Performance metrics
- Incentive systems wholesale

Those changes belong later—after design intent is clear.

How Alex Approached the Transition

At NorthRiver Foods, Alex resisted pressure to act quickly.

Instead of launching fixes, he focused on shared understanding.

He revisited friction maps with leaders and asked:

- Where is friction costing us the most today?
- What would happen if this friction remained for six more months?
- What design change would reduce the most drag with the least disruption?

These questions slowed the organization just enough to move deliberately.

Why Leaders Must Normalize Discomfort

Designing friction out of systems feels uncomfortable—not because it is wrong, but because it removes familiar safeguards.

Leaders must normalize this discomfort.

That means:

- Acknowledging uncertainty
- Making trade-offs explicit
- Accepting temporary instability

When leaders pretend redesign is painless, trust erodes.

When leaders name the discomfort and proceed anyway, confidence grows.

The Principle of Minimum Effective Design

One of the most useful principles at this stage is minimum effective design.

The question is not:
"What is the perfect system?"

It is:
"What is the smallest design change that meaningfully reduces friction?"

Small, intentional changes:

- Are easier to absorb
- Reveal system responses
- Prevent unintended consequences

Transformation gains durability through iteration—not overhaul.

Why Patience Accelerates Change

Ironically, patience at this stage accelerates progress.

When leaders:

- Sequence design
- Protect stability where needed
- Signal support for decision-makers

People move faster.

Hesitation decreases not because pressure increases, but because the system begins to feel safer.

What This Chapter Enables

By the end of this chapter, leaders should not feel compelled to act immediately.

They should feel prepared to act intentionally.

The next chapters will describe:

- What to design
- How to design it
- How to sustain it

But only after this moment of restraint.

Reflection for the Reader

If friction is now visible, ask:

- What is the smallest change that would reduce the most drag?
- What stability must be preserved right now?
- Where is patience more valuable than urgency?

Do not rush to fix.
Prepare to design deliberately.

Chapter 7 Summary

- Awareness alone does not change systems.
- Alignment without enablement stalls progress.
- Familiar friction often feels safer than redesign.
- Systems change through structure, not intention.
- Overcorrection creates new friction.
- Design must be sequenced deliberately.
- Minimum effective design outperforms overhaul.

Questions for the Reader

- Where have we agreed—but not redesigned?
- Which controls exist for past problems that no longer apply?
- What behaviors does our system still reward?
- Where are leaders expecting behavior change without structural change?
- Where would small structural adjustments outperform large initiatives?

These are not communication questions.
They are design leadership questions.

Resist the urge to overhaul. Sequence deliberately.

What Comes Next

Diagnosis without redesign sustains friction. The shift from awareness to action requires disciplined design—not broad mandates.

In the next chapter, Designing Systems That Make the Right Behavior Easy, we move from identifying friction to building operating principles that shape decisions, ownership, information, and risk by design.

This is where transformation stops being heavy—and starts becoming deliberate.

Chapter 8: Designing Systems That Make the Right Behavior Easy

Why Behavior Follows Design

Organizations often try to change behavior directly.

They communicate more.
They train harder.
They reinforce expectations.

When those efforts fail, leaders conclude that people are resistant, disengaged, or unwilling to change.

But in Operational Excellence, we know a different truth:

Behavior is an output of system design.

If the right behavior feels difficult, risky, or slow, the system is working against it.

Designing systems that make the right behavior easy is not motivational work.
It is structural work.

From Removing Friction to Designing Flow

Up to this point, the book has focused on identifying friction—where transformation slows and why.

Designing flow is the next step.

Flow does not mean speed at all costs.
It means work moves with clarity, confidence, and minimal drag.

Flow exists when:

- Ownership is clear
- Decisions are timely
- Information supports action
- Risk is shared appropriately

These conditions do not happen accidentally.

They must be designed.

Month Twelve: A Shift in How Work Moved

At NorthRiver Foods, Alex noticed something subtle but important as design conversations replaced diagnostic ones.

Leaders stopped asking:
"What do we need approval for?"

They started asking:
"What should be easy—and what should be hard?"

This reframing changed everything.

Instead of optimizing controls, the organization began optimizing flow.

Design Principle 1: Make Ownership Obvious

When ownership is clear, hesitation drops.

This does not require new roles.
It requires explicit design.

Effective ownership design:

- Names owners for outcomes, not activities
- Aligns authority with accountability
- Extends beyond approval into sustainment

When ownership is obvious, people act without seeking permission.

Design Principle 2: Separate Decisions From Approvals

Approvals should protect the system.
Decisions should move the work.

When these are confused, friction grows.

Designing decision flow means:

- Defining who decides
- Clarifying which decisions are reversible
- Establishing escalation paths in advance
- Limiting approvals to true risk boundaries

When decisions are designed, alignment becomes faster—and quieter.

Design Principle 3: Design Information for the Moment of Action

Information should arrive where decisions are made—not where it is convenient to report.

This requires discipline.

Effective information design:

- Starts with the decision, not the metric
- Defines expected responses to signals
- Assigns ownership for interpretation
- Reduces debate instead of fueling it

Information that does not support action is friction.

Design Principle 4: Share Risk Where You Expect Action

People move faster when they know they will not be left alone.

Designing shared risk means:

- Making post-decision support explicit
- Normalizing learning—not just success

- Aligning incentives with outcomes
- Reducing personal exposure for system-level bets

When risk is shared, confidence increases naturally.

What These Design Principles Look Like in Practice

Design principles are not rules to be followed mechanically. They are lenses leaders use to make better trade-offs—especially when pressure is high and information is incomplete.

In practice, applying these principles rarely looks dramatic. It looks deliberate.

At NorthRiver Foods, once the principles were articulated, leaders did not rush to redesign everything. Instead, they began using the principles as a filter—a way to evaluate choices already in front of them.

Using Principles to Decide What Not to Fix

One of the first changes leaders noticed was what they stopped doing.

They stopped trying to eliminate all friction.
They stopped chasing perfect alignment.
They stopped over-optimizing controls.

Instead, they asked simpler questions:

- Does this friction block decisions or just create discomfort?
- Are we adding clarity—or adding steps?
- Are we protecting the system, or protecting ourselves?

When a proposed change did not clearly reduce friction, it was deferred—even if it sounded reasonable.

Restraint became a form of discipline.

How Principles Guide Trade-Offs Under Pressure

Design principles proved most valuable when trade-offs were unavoidable.

For example:

- When speed conflicted with certainty, leaders asked whether the decision was reversible.
- When risk felt personal, they asked whether support was explicit.
- When information was debated, they asked what decision the data was meant to serve.

The principles did not remove disagreement.
They changed the basis of disagreement.

Conversations shifted from opinion to design.

Making the Right Behavior Easier—Quietly

As principles were applied consistently, behavior began to change without mandates.

Leaders:

- Acted without waiting for permission

- Escalated issues earlier
- Accepted temporary imperfection
- Trusted decisions to hold

Not because expectations were reinforced—but because the system made those behaviors safer.

The most noticeable change was not speed.
It was confidence.

What Leaders Learned to Leave Imperfect

Perhaps the most counterintuitive outcome was leaders' willingness to leave some things unresolved.

Not every role was perfectly defined.
Not every decision path was optimized.
Not every metric was redesigned.

And that was intentional.

The principles helped leaders distinguish between:

- Friction that mattered
- Friction that was tolerable

This prevented redesign from becoming disruption.

Principles as a Shared Reference Point

Over time, the principles became part of the organization's working language.

Leaders referenced them informally:

- "This feels like decision friction."
- "We're making this harder than it needs to be."
- "Risk isn't aligned here."

The principles did not replace governance.
They improved how governance was used.

Why This Matters Before Moving to Tools

This stage—using principles before tools—is critical.

Organizations that jump straight to frameworks often:

- Over-design
- Create dependency
- Lose adaptability

By contrast, principles:

- Scale without documentation
- Adapt to new situations
- Survive leadership changes

They create design judgment, not compliance.

Why Design Principles Matter More Than Rules

Rules are brittle.

They address specific scenarios and fail when conditions change.

Design principles endure.

They guide decisions when:

- Information is incomplete
- Trade-offs are necessary
- New situations emerge

At NorthRiver, Alex saw leaders begin referencing principles instead of policies.

This was the turning point.

Design Is About Constraints, Not Freedom

One of the most counterintuitive lessons in system design is this:

Constraints create speed.

When boundaries are clear:

- Decisions accelerate
- Conflict decreases
- Work moves predictably

Ambiguity feels flexible—but it creates friction.

Designing constraints is not restrictive.
It is enabling.

Why Leaders Avoid Design Work

Design work is uncomfortable.

It forces leaders to:

- Remove legacy controls

- Accept new risk distributions
- Let go of familiar approval paths
- Commit to trade-offs

Compared to launching initiatives, design work feels slow and abstract.

But it is the only work that changes how the organization behaves under pressure.

From Initiative-Led to System-Led Transformation

As NorthRiver shifted from initiatives to system design, transformation felt lighter.

Not easier—but steadier.

Initiatives no longer depended on heroics.
Progress no longer required constant reinforcement.

The system began producing the behavior leaders wanted—without asking for it.

Designing for the Organization You Have

Designing systems is not about perfection.

It is about fit.

Effective design:

- Reflects organizational maturity
- Respects existing constraints

- Evolves incrementally
- Prioritizes clarity over elegance

Transformation succeeds when design matches reality—not aspiration.

Chapter 8 Summary

- Behavior follows system design.
- Flow must be intentionally structured.
- Principles outperform rules.
- Design reduces the need for control.
- Shared risk accelerates action.

Questions for the Reader

- What behaviors should be easy—but aren't?
- Where does the system require workarounds?
- Which decisions feel heavier than they should?
- Where could clarity replace control?
- Where are rules compensating for missing principles?

These are not culture questions.
They are design questions.

Make the right behavior easier than the workaround.

What Comes Next

Design principles provide direction, but they must be translated into operational practice to remove friction consistently.

In the next chapter, From Principles to Practice, we examine how to embed these principles into daily execution—so transformation becomes sustained, not episodic.

Chapter 9: From Principles to Practice

When Principles Meet Reality

Design principles are easy to agree with in theory.

Make ownership clear.
Separate decisions from approvals.
Design information for action.
Share risk where you expect movement.

Alex learned quickly that agreement did not mean readiness.

When these principles moved from discussion into practice, they collided with legacy structures, informal power, and long-standing habits. This was not resistance. It was reality.

Transformation doesn't stall because principles are wrong.
It stalls because systems are already optimized for yesterday's problems.

The First Conversation: Ownership Without Illusion

Alex started where friction was most visible: ownership.

In a working session with an operations leader, the conversation turned to a stalled initiative that had been labeled "cross-functional."

Alex asked a simple question:

"Who owns the outcome if this doesn't deliver?"

The room went quiet.

Several leaders described their involvement.
None described accountability.

Together, they drafted a charter—not to document the work, but to define responsibility.

The charter did three things:

- Named a single outcome owner
- Clarified decision authority
- Extended ownership beyond implementation

No new role was created.
No hierarchy changed.

But something shifted.

The work now had a center of gravity.

Charters as Friction-Reduction Tools

Alex was careful about how charters were framed.

These were not governance artifacts for leadership decks.
They were operating agreements.

Each charter answered:

- What problem are we solving?
- Who owns the outcome?
- What decisions does that owner control?
- How will conflicts be resolved?

By design, charters removed ambiguity.

And ambiguity was the largest source of friction.

Designing Decision Flow With IT and Finance

The most difficult conversations came next.

Alex facilitated a session with IT and Finance leaders to examine decision flow. The goal was not speed—it was clarity.

They mapped a recent initiative end to end:

- Where approvals occurred
- Where decisions stalled
- Where escalation felt unsafe

What they saw surprised them.

Approvals had multiplied over time, but decision authority had not moved with them.

Finance wanted predictability.
IT wanted stability.
Operations wanted momentum.

Each function was protecting what it owned.

No one was protecting flow.

Replacing Approval Chains With Decision Design

Instead of debating which approvals to remove, Alex reframed the discussion.

"What decisions do we want leaders to make confidently—and which should escalate?"

That question changed the tone.

They categorized decisions:

- Reversible vs irreversible
- Local vs enterprise-impacting
- Operational vs architectural

From there, decision rights were explicitly assigned.

Approvals didn't disappear.
They became purposeful.

Information Redesign Happens in Meetings, Not Dashboards

Alex tackled information friction next.

Rather than launching another analytics project, he attended operational reviews and asked one disruptive question:

"What decision is this data meant to support?"

Often, no one could answer.

Metrics had been added over time without removing old ones.
Dashboards reflected availability—not usefulness.

Working with OpX and site leaders, Alex led a redesign focused on decisions:

- What should a leader do differently if this metric changes?
- Who owns that decision?
- How fast does the signal need to arrive?

The result was not more data.

It was less—but better—information.

When Metrics Finally Drove Action

As information was redesigned, something changed.

Debate decreased.
Confidence increased.
Leaders acted sooner.

Not because data improved—but because information was finally aligned to decisions.

The system no longer asked leaders to interpret signals alone.

The Hardest Principle: Designing Shared Risk

Risk conversations were the most sensitive.

In a leadership discussion about scaling a successful pilot, hesitation surfaced again.

Alex didn't push for urgency.

He asked:
"What support exists after approval if this pilot faces challenges?"

The answer was unclear.

That was the friction.

Together, leadership defined explicit post-decision support:

- Escalation paths
- Resource commitments
- Learning expectations

Failure was no longer silent or personal.

It was shared and managed.

What Changed—and What Didn't

Not everything improved immediately.

Some approvals remained.
Some decisions still escalated.
Some leaders needed time to adjust.

But one thing was unmistakable.

Transformation felt lighter.

Work moved with fewer handoffs.
Decisions stuck.
Pilots transitioned to scale more often.

The system no longer worked against people trying to do the right thing.

Why This Worked

Alex didn't implement tools.
He didn't restructure the organization.
He didn't mandate behavior.

He helped leaders redesign how the system worked.

That is the difference between transformation that feels forced—and transformation that feels natural.

Charters Are a Mechanism—Not the Method

It is important to pause here and clarify something.

Charters are not required for removing friction.
Clarity is.

Charters appear in this chapter because they are one effective way to make ownership, decision rights, information flow, and risk explicit. They are a mechanism—not the method itself.

The real work is not creating documents.
The real work is designing systems that leave no ambiguity about:

- Who owns outcomes
- Who decides and when
- What information supports those decisions
- How leaders are supported when risk is introduced

If those questions are answered clearly and consistently, friction decreases—regardless of the tool used to capture them.

Other Ways Organizations Achieve the Same Outcome

Different organizations use different mechanisms to express the same design intent. In practice, the following approaches often serve the same purpose as charters when designed well:

- Decision-rights frameworks that explicitly define who decides, who contributes, and who must be informed

- Operating agreements between functions that clarify ownership, escalation paths, and expectations
- Governance models that separate approval from decision authority
- Leadership standard work that embeds decision-making and review into operating rhythms
- Team working agreements that define responsibility and accountability at the execution level
- Standing decision forums where recurring decisions are intentionally owned and supported

The form matters far less than the outcome.

What matters is that the system makes responsibility visible, decisions predictable, and risk shared.

Why Charters Are Often a Useful Starting Point

Charters are often effective not because they are sophisticated, but because they are simple.

They force conversations that organizations frequently avoid.
They make assumptions explicit.
They create a shared reference point when priorities conflict.

In organizations with low or uneven maturity—particularly where transformation spans operations, IT, finance, and OpX—charters can serve as a practical entry point for system design.

But they are not the only path.

The Design Test

Regardless of the mechanism used, leaders should be able to answer these questions without hesitation:

- Who owns this outcome end to end?
- Who can decide—and who supports that decision?
- What information is required to act?
- How is risk shared after commitment?

If the system answers these questions clearly, friction falls.

If it does not, no artifact—charter or otherwise—will compensate.

Chapter 9 Summary

- Principles require operating mechanisms.
- Charters clarify ownership.
- Decision design replaces approval overload.
- Information must serve action.
- Shared risk unlocks movement.
- Design must be embedded in daily execution.

Questions for the Reader

- Where are principles agreed but not embedded?
- Which charters would remove ambiguity immediately?
- Where are approvals substituting for decision design?
- How is risk shared after decisions are made?
- Which mechanisms exist—but are not actively used?

These are not execution questions.
They are leadership design questions.

Embed structure before demanding performance.

What Comes Next

Design without sustainment fades. Systems revert when structure is not reinforced through rhythm and leadership behavior.

In the final chapter, Making Transformation Stick, we examine how organizations prevent friction from returning—by embedding design into operating rhythms, leadership practice, and continuous improvement systems.

Chapter 10: Making Transformation Stick

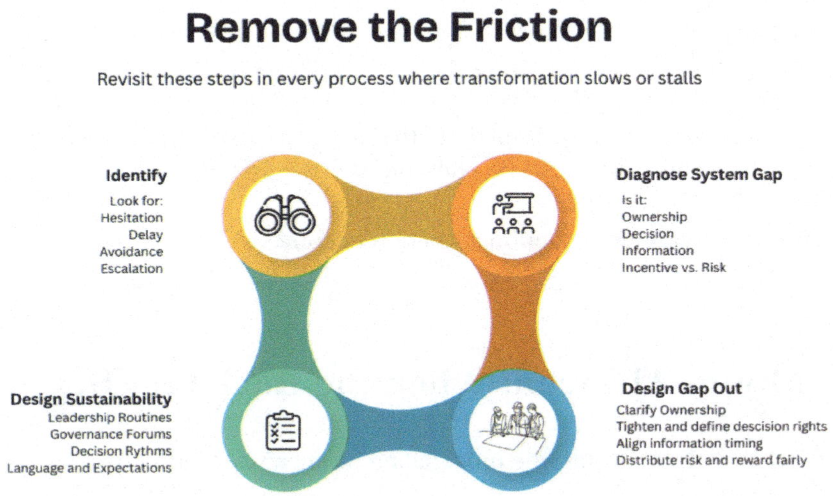

Why Friction Has a Way of Coming Back

Even well-designed systems drift.

New leaders arrive.
Priorities shift.
Pressure increases.
Shortcuts reappear.

Without intentional sustainment, friction slowly returns, not because people forget, but because systems are rarely designed to self-correct.

Making transformation stick is not about maintaining momentum.
It is about embedding design into how the organization operates every day.

Sustainment Is a System, Not a Phase

Many organizations treat sustainment as the final step.

Implement.
Stabilize.
Move on.

But sustainment is not something that happens after transformation.
It is something that must be built into the system itself.

When sustainment is optional, friction becomes inevitable.

Month Thirteen: When the Real Test Begins

At NorthRiver Foods, the most telling moment came after visible progress.

Ownership was clearer.
Decisions moved faster.
Information supported action.
Risk felt shared.

And then… pressure returned.

A new initiative emerged.
Timelines tightened.
Old approval paths resurfaced quietly.

Alex didn't intervene immediately.

He observed.

This was the real test—not of intent, but of design.

Embedding Design Into Operating Rhythms

What ultimately prevented regression was not vigilance—it was rhythm.

The organization began embedding design principles into routines that already existed:

- Leadership reviews included explicit decision ownership
- Initiative updates required clarity on risk and support
- Governance forums revisited decision rights, not just status
- OpX routines surfaced friction before it compounded

These were not new meetings.

They were existing rhythms redesigned.

Why Leadership Standard Work Matters

Sustainment lives in what leaders do repeatedly.

Not what they announce.
Not what they sponsor.

But what they consistently ask for and reinforce.

Leadership standard work became a quiet stabilizer:

- Asking who owns outcomes
- Clarifying which decisions were being made
- Challenging approval-heavy thinking
- Surfacing friction as data

Over time, this changed expectations.

Friction was no longer hidden.
It was discussed early.

From Continuous Improvement to Continuous Design

CI organizations are good at improving processes.

But transformation requires something adjacent—and often missing:

Continuous design.

This means periodically asking:

- Does this system still support the behavior we want?
- Where has friction quietly reappeared?
- What decisions have outgrown their original design?

Design is not a one-time event.
It is an ongoing responsibility.

Why Metrics Alone Cannot Sustain Change

Metrics tell you what is happening.

They rarely tell you why.

Sustainment relies less on performance metrics and more on structural indicators, such as:

- Are decisions being revisited?
- Is ownership still clear?
- Are approvals increasing?
- Is risk quietly becoming personal again?

These signals are qualitative—but powerful.

They are early warnings of friction.

Alex Learns to Measure What Matters

Rather than tracking adoption directly, Alex and OpX tracked indicators of friction:

- Time to decision
- Number of approval layers
- Escalation frequency
- Reopened decisions

These were not KPIs.
They were design health indicators.

When these drifted, leaders intervened—not with pressure, but with redesign.

Why Transformation Fails When It Becomes "Someone's Job"

One of the most common sustainment failures occurs when transformation becomes owned by a function instead of the system.

When:

- OpX owns transformation
- IT owns digital
- Operations owns execution

Friction returns at the seams.

Sustainment works when leaders across the system own design, not just outcomes.

The Quiet Power of Shared Language

Perhaps the most durable change at NorthRiver was linguistic.

Leaders stopped saying:

- "People are resisting"
- "We need alignment"
- "This is a culture issue"

They started saying:

- "Where is the friction?"
- "What's unclear by design?"
- "What decision is this really about?"

Language shapes attention.

Attention shapes behavior.

What Making Transformation Stick Really Means

Making transformation stick does not mean:

- No friction
- No conflict
- No hesitation

It means:

- Friction is visible
- Design is intentional
- Decisions are supported
- Risk is shared
- Systems evolve deliberately

Transformation becomes how the organization works, not what it is trying to do.

Final Summary

Across this book, one idea has remained constant:

Friction in transformation is a design failure—not a people problem.

When leaders stop managing behavior and start designing systems:

- Progress accelerates
- Trust increases
- Change becomes sustainable

This is not about perfection.
It is about responsibility.

A Final Reflection for the Reader

If transformation feels heavy, pause.

Do not ask:

- Who is resisting?
- Why is this hard?
- How do we push harder?

Ask instead:

- Where is the friction?
- What is unclear by design?
- What needs to be redesigned?

That question changes everything.

Closing Thought

The goal of transformation is not speed.
It is confidence.

Confidence that the system will support people when pressure rises.
Confidence that decisions will hold.
Confidence that learning will be shared.

When systems are designed well, people don't need to be convinced.

They move.

Appendix A: Running a Friction Mapping Session

Purpose of a Friction Mapping Session

A Friction Mapping Session is not a problem-solving meeting.

Its purpose is to:

- Surface where work slows
- Make system design visible
- Separate friction from blame
- Create shared understanding before redesign begins

This session is about seeing, not fixing.

Organizations that rush to solutions before friction is clearly understood often redesign the wrong thing.

When to Run a Friction Mapping Session

A Friction Mapping Session is most valuable when:

- Transformation efforts feel heavy or slow
- Initiatives stall without obvious resistance
- Decisions are revisited repeatedly
- Leaders sense hesitation but can't name why

It is particularly effective:

- Early in transformation efforts
- After repeated pilot failures
- When cross-functional tension is present

Who Should Be in the Room

The quality of a Friction Mapping Session depends on representation—not hierarchy.

Include:

- Leaders who own outcomes
- Leaders who make decisions
- Leaders who absorb risk
- One facilitator (often OpX, CI, or Transformation)

Avoid:

- Large audiences
- Observers without responsibility
- People sent "to listen" but not act

The session works best with 6–10 participants.

How Long It Takes

A typical session requires:

- 90 minutes for first-time groups
- 60 minutes once the organization is familiar

Shorter sessions encourage surface-level discussion.
Longer sessions tend to drift into solutioning.

How the Session Is Framed

The facilitator should establish three rules at the start:

1. No solutions
2. No blame
3. No defending the system

The goal is to observe how the system behaves—not to justify it.

The Core Questions

The session follows four disciplined questions, in order:

1. What work or decision are we trying to move?
   (Be specific—avoid abstractions.)
2. Where does it slow or stop?
   (Name moments, not people.)
3. What design gap exists at that point?
   (Ownership, decision, information, or risk.)
4. How does that gap show up in behavior?
   (Hesitation, delay, escalation, avoidance.)

This structure keeps the discussion objective and grounded.

What the Facilitator Should Listen For

Effective facilitators listen for patterns, not opinions.

Signals of friction include:

- Repeated references to "alignment"
- Vague ownership language
- Hesitation around authority
- Risk framed as personal exposure
- Data debates instead of decisions

These signals are data.

What to Capture (and What Not To)

Capture:

- Friction points
- Design gaps
- Observable behaviors

Do not capture:

- Action items
- Solutions
- Names tied to problems
- Commitments

The output should feel unfinished—because it is.

What a Good Session Output Looks Like

A strong Friction Mapping Session produces:

- A shared understanding of where friction exists
- Agreement on which friction matters most
- Reduced tension and defensiveness
- Readiness for design—not urgency to fix

If the group leaves feeling compelled to act immediately, the session moved too fast.

What to Do After the Session

The next step is not implementation.

The next step is:

- Reflection
- Prioritization
- Sequencing

Leaders should ask:

- Which friction creates the most drag today?
- What design change would reduce friction without destabilizing the system?
- What must remain stable while change occurs?

These questions prepare the organization for intentional design.

Common Mistakes to Avoid

- Turning the session into a brainstorming meeting
- Allowing solutions to dominate discussion
- Over-representing one function
- Using the output as performance evidence

A Friction Map is a leadership tool, not an audit artifact.

Why This Session Works

Friction Mapping works because it:

- Makes invisible system behavior visible
- Removes blame from the conversation
- Builds shared language
- Creates readiness for change

Most importantly, it slows the organization just enough to move forward deliberately.

Final Thought

A Friction Mapping Session does not solve problems.

It creates the conditions for solving the right problems—in the right order.

That is the work of leadership.

Made in the USA
Monee, IL
03 November 2026

67385834R00079